LUKE BOLINGER

The Basics of 3D Printing

A beginner-friendly guide to 3D printing, covering basics, advanced tips, and future insights.

First edition

This book was professionally typeset on Reedsy.
Find out more at reedsy.com

Contents

1

Introduction

Brief History of 3D Printing

The concept of 3D printing, or additive manufacturing, emerged in the 1980s with the invention of stereolithography (SLA) by Chuck Hull. This technology laid the foundation for the development of 3D printing as we know it today. Initially, 3D printing was primarily used for rapid prototyping in industrial settings due to the high cost of printers and materials. Over the years, advancements in technology have significantly reduced these costs, making 3D printing accessible to hobbyists, educators, and small businesses.

The Basic Principles of 3D Printing

At its core, 3D printing involves creating physical objects from digital models by adding material layer by layer. This additive process contrasts with traditional subtractive manufacturing methods, which remove material to create shapes. The ability to directly translate digital designs into physical objects revolutionizes how we think about manufacturing, enabling the creation of complex geometries that would be difficult or impossible to achieve with conventional methods.

Importance and Applications in Various Fields

3D printing has found applications in a wide range of fields, from aerospace and automotive to healthcare and fashion. In aerospace, companies use 3D printing to produce lightweight, complex components that reduce fuel consumption. In healthcare, 3D printing is used to create custom prosthetics, dental implants, and even bio-printed tissues. The technology's versatility also enables entrepreneurs and designers to rapidly prototype new products, customize items, and even produce end-use parts on demand.

2

Understanding the Basics of 3D Printing

Types of 3D Printing Technologies

3D printing encompasses several technologies, each with its unique mechanisms and applications.

- **Fused Deposition Modeling (FDM):** FDM is the most common 3D printing technology, known for its affordability and ease of use. It works by extruding thermoplastic filaments through a heated nozzle, layer by layer, to build an object. FDM printers are popular among hobbyists and in education for their simplicity and the wide range of available materials.
- **Stereolithography (SLA):** SLA, the first 3D printing technology developed, uses an ultraviolet laser to cure and solidify liquid resin into the desired shapes. SLA printers are known for producing high-resolution objects with smooth surfaces, making them ideal for applications requiring fine details, such as jewelry and dental models.
- **Selective Laser Sintering (SLS):** SLS technology uses a laser to sinter

powdered material, typically nylon or polyamide, to form solid objects. Unlike FDM and SLA, SLS does not require support structures as unused powder supports the object during printing. This feature makes SLS suitable for complex geometries and functional parts.

Materials Used in 3D Printing

The choice of materials in 3D printing is vast and varies depending on the technology used.

- **Plastics:** The most commonly used materials in 3D printing include PLA (Polylactic Acid) and ABS (Acrylonitrile Butadiene Styrene). PLA is biodegradable and user-friendly, making it a favorite for beginners, while ABS is known for its strength and durability, suitable for more functional applications.
- **Resins:** Used in SLA printing, resins can be formulated to achieve various properties, such as flexibility, high-temperature resistance, and transparency. This versatility makes resin printing popular for detailed models and prototypes.
- **Metals:** Metal 3D printing is primarily used in industrial applications for creating complex metal parts directly from digital models. Technologies like Direct Metal Laser Sintering (DMLS) enable the production of components in aerospace, automotive, and medical industries.

Basic Components of a 3D Printer

Understanding the key components of a 3D printer is essential for operating and troubleshooting.

- **Frame:** The frame provides the structure of the printer, ensuring stability during printing.

- **Extruder:** In FDM printers, the extruder is responsible for feeding filament through a heated nozzle to melt and deposit the material.
- **Print Bed:** The surface onto which the 3D object is printed. Some print beds are heated to prevent the warping of materials like ABS.

Control Panel: Many printers feature a control panel or interface for navigating settings and starting prints. More advanced printers may offer connectivity options for remote management.

3

Getting Started with 3D Printing

Choosing Your First 3D Printer

The selection of a 3D printer is influenced by several factors, including budget, intended use, and the level of detail required for prints. Here are key considerations:

- **Budget:** Determine how much you're willing to spend. FDM printers are generally more affordable and offer a good starting point for beginners.
- **Print Quality:** For projects requiring high detail, such as miniature models or intricate jewelry, an SLA printer might be more appropriate despite its higher cost.
- **Ease of Use:** Look for printers with supportive communities, extensive documentation, and user-friendly interfaces, especially if you're new to 3D printing.
- **Material Compatibility:** Consider what materials you plan to use. If you're interested in printing with a variety of materials, ensure the printer can accommodate them.

Setting Up Your 3D Printing Workspace

A well-organized and safe workspace is essential for efficient and enjoyable 3D printing. Here's how to set it up:

- **Ventilation:** Ensure good airflow, especially when printing with materials that emit fumes, like ABS.
- **Storage:** Keep your filaments and resins in a dry, cool place to prevent degradation. Humidity can significantly affect material properties.
- **Workspace Layout:** Arrange your tools, printer, and materials for easy access. A clutter-free environment reduces the risk of accidents and improves printing efficiency.

Safety Measures and Best Practices

3D printing involves high temperatures and, in some cases, hazardous materials. Here are important safety tips:

- **Wear Protective Gear:** Use gloves when handling hot components or post-processing prints. Safety glasses are recommended when cutting or sanding.
- **Understand Your Printer:** Familiarize yourself with your printer's safety features and how to shut it down in case of an emergency.
- **Material Safety:** Always use materials as recommended by your printer manufacturer. Pay attention to the material safety data sheets (MSDS) for any specific health risks.

With your printer selected, workspace prepared, and safety measures in place, you're now ready to embark on your 3D printing journey. The joy of bringing digital creations to life awaits, along with the satisfaction of solving practical problems with your prints.

4

Designing for 3D Printing

Introduction to 3D Modeling Software

Choosing the right 3D modeling software is the first step in the design process. There are several options available, each with its strengths and learning curves.

- **Tinkercad:** Ideal for beginners, Tinkercad is a free, web-based application that offers an intuitive interface for creating simple 3D models. It's perfect for educational purposes and basic projects.
- **Fusion 360:** Aimed at more experienced users, Fusion 360 is a powerful tool for precision engineering and design. It offers advanced features like parametric tools and simulation capabilities, making it suitable for professional-grade projects.
- **Blender:** Blender is a free, open-source software known for its versatility in 3D modeling, animation, and rendering. While it has a steeper learning curve, Blender is incredibly powerful for those willing to master its complexities.

Designing Your First 3D Model

When designing your first 3D model, start with a simple project to familiarize yourself with the software's tools and features. Here are some tips for success:

- **Begin with Tutorials:** Most 3D modeling software comes with tutorials. Starting with these can help you understand the basics of model creation.
- **Think in Layers:** Remember that your design will be printed layer by layer. Designing with this in mind can help avoid issues such as overhangs that require support structures.
- **Use Reference Materials:** If you're trying to replicate an existing object or bring a concept to life, reference images can be invaluable for accuracy and detail.

Preparing Your Model for Printing (Slicing)

Once you've created a 3D model, it must be prepared for printing through a process called slicing. Slicing software converts your 3D model into instructions that the printer can understand, including details about layers, supports, and print speed.

- **Choosing Slicing Software:** Many printers come with their own slicing software, but there are universal options like Ultimaker, Cura, and PrusaSlicer that offer flexibility and advanced features.
- **Orientation and Supports:** Properly orienting your model and adding supports can significantly impact the success of your print. Orientation affects the strength and appearance of the final product, while supports are necessary for overhangs and complex structures.
- **Layer Height and Fill Density:** These settings can be adjusted based

on the desired strength and finish of your object. A lower layer height increases detail but extends print time, while fill density impacts the object's weight and solidity.

By mastering 3D modeling and slicing, you can transform your ideas into printable designs. This process of designing, iterating, and printing is at the heart of 3D printing's appeal, offering endless possibilities for customization and innovation.

5

Advanced Techniques and Materials

Exploring Advanced 3D Printing Materials

Beyond the common plastics like PLA and ABS, the 3D printing world offers a variety of advanced materials, each with unique properties and applications:

- **Flexible Filaments (TPU/TPE):** These materials can stretch and bend, making them ideal for printing parts like flexible hinges, phone cases, or wearable items.
- **Composite Filaments:** Infused with materials such as carbon fiber, wood, or metal, composite filaments can mimic the appearance and properties of the infused material, adding strength or aesthetic qualities to prints.
- **High-Temperature and Chemical-Resistant Plastics:** For applications requiring durability under harsh conditions, materials like Polycarbonate (PC) and Polypropylene (PP) offer superior toughness and resistance.

Introduction to Support Structures and Dual Extrusion

- **Support Structures:** Complex prints with overhangs or suspended parts often require support structures to prevent collapse during printing. Designing efficient supports that are easy to remove while providing adequate support is a skill that improves with experience.
- **Dual Extrusion:** Printers equipped with dual extruders can print with two different materials or colors simultaneously. This capability is particularly useful for printing supports with soluble filament, which can be dissolved after printing, leaving behind a clean, high-quality print.

Post-Processing Techniques for 3D Printed Objects

The appearance and functionality of 3D-printed objects can be significantly enhanced through various post-processing techniques:

- **Sanding and Filling:** For a smooth finish, sanding the print to remove layer lines, followed by filling any gaps with putty, can prepare the surface for painting or other finishes.
- **Painting:** Acrylic paints are commonly used to add color and detail to prints. Applying a primer before painting can help achieve an even finish.
- **Chemical Smoothing:** Certain materials, like ABS, can be smoothed with chemicals such as acetone vapor, which melts the outer layer, resulting in a glossy surface.

Advanced Design Considerations

When designing for advanced applications, consider the following:

- **Material Properties:** Choose a material that best matches the mechanical and aesthetic requirements of your project.
- **Print Orientation and Layout:** Optimize the orientation and layout of your print for the best strength, surface finish, and material efficiency.
- **Tolerance and Fit:** For functional parts that fit together, account for the printer's accuracy and material shrinkage to ensure a proper fit.

Exploring advanced materials and techniques opens up new possibilities for what you can create with a 3D printer. From functional mechanical components to artistic models, the ability to select the right material and apply suitable post-processing methods can significantly impact the outcome of your projects.

6

Troubleshooting and Maintenance

Common 3D Printing Problems and Solutions

- **Warping:** Warping occurs when the first layers of the print do not adhere well to the build plate, causing the corners of the print to lift. Solutions include using a heated bed, applying adhesives (like glue stick or hairspray) to the build plate, and ensuring the printing environment is free from drafts.
- **Stringing:** Stringing happens when small strings of plastic are left between different parts of the print, caused by oozing filament as the extruder moves. To reduce stringing, adjust retraction settings in your slicer software, which pulls the filament back when the extruder moves across open spaces.
- **Clogging:** A clogged nozzle can disrupt the flow of filament, leading to incomplete prints or extrusion problems. Regularly cleaning the nozzle, ensuring the filament is free from contaminants, and using the correct printing temperature for the material can prevent clogging.

Maintaining Your 3D Printer for Longevity and Performance

- **Regular Lubrication:** Moving parts, such as rods and bearings, should be lubricated regularly with appropriate oils or greases to ensure smooth operation.
- **Keeping Firmware Updated:** Manufacturers often release firmware updates that improve functionality or address known issues. Keeping your printer's firmware up to date can enhance performance and stability.
- **Calibrating Your Printer:** Periodically recalibrate your printer's bed leveling, extruder, and other settings to ensure accuracy. Calibration is key to achieving consistent print quality.
- **Cleaning:** Keep the printer clean from dust and filament residues. A clean printer is less likely to encounter operational issues.

Creating a Maintenance Schedule

Developing a regular maintenance schedule helps in identifying potential issues before they lead to failures. Include checks for wear and tear on components, updating software, and cleaning as part of your routine. Consistent maintenance not only extends the life of your printer but also ensures that it remains a reliable tool for your 3D printing projects.

Learning from Failures

3D printing, like any skill, involves a learning curve, and failures are part of the process. Each failed print provides valuable insights into what can be improved, whether it's adjusting design parameters, changing materials, or fine-tuning printer settings. Engaging with the community through forums and social media can also offer solutions and support from fellow enthusiasts.

7

The Future of 3D Printing

As we look toward the horizon of 3D printing, it's clear that this technology is not just a tool for creating physical objects but a transformative force across industries. This final section explores emerging trends, the potential impact of 3D printing on various sectors, and how advancements in this technology might shape our future.

Emerging Trends in 3D Printing

- **Metal 3D Printing:** Once prohibitively expensive and complex, metal 3D printing is becoming more accessible, opening up new possibilities in aerospace, automotive, and medical industries for producing durable, high-strength parts.
- **Bioprinting:** 3D bioprinting, the process of creating cell patterns in a confined space using bio-inks, is advancing towards printing functional human tissues and organs. This could revolutionize transplant medicine and pharmaceutical testing.
- **Sustainability:** The drive towards sustainable manufacturing practices is pushing the development of eco-friendly materials and processes in 3D printing. This includes recycling materials and

reducing waste in production.

Impact of 3D Printing on Industries

- **Manufacturing:** 3D printing is leading to a shift towards on-demand production and mass customization, reducing inventory costs and waste. It allows for the rapid prototyping and production of parts, significantly speeding up the development process.
- **Medicine:** Beyond prosthetics and dental applications, 3D printing is making strides in creating patient-specific implants and even bioprinting tissues, offering personalized treatment options and potentially reducing the need for organ donors.
- **Construction:** The use of 3D printing in construction is emerging, with the technology being used to create components off-site or even build entire structures. This can lead to reduced material waste and labor costs, and potentially make building in remote or resource-limited areas easier.

The Future of 3D Printing Technologies

The future of 3D printing lies in continuous innovation and overcoming current limitations, such as print speed, material capabilities, and size constraints. Developments in AI and machine learning could lead to smarter printers that optimize themselves for different materials and designs, further democratizing manufacturing.

8

Conclusion

3D printing stands at the forefront of technological innovation, promising to change how we create, innovate, and produce goods. From individual hobbyists to large-scale industries, the applications and implications of 3D printing are vast and expanding. As technology advances, it will continue to break barriers, offering new solutions to old problems and opening up unprecedented opportunities for creativity and innovation.

Resources

Chat-GPT 4.0 - 2024